Contents

Foreword

Glossary

Foreword

Water Pollution; provides a unique insight into the problems our planet faces in terms of water quality and quantity, and what to do about it. This is the only books expressed comprehensive and interdisciplinary focus to hydrological understanding with the multidimensional approach.

This book made of 06 years consistently research on water resources, makes it ideal source for students, teachers, industrialist, water experts and environmentalists.

This book provides an essential guide to researchers, it offers: various aspects of water; on the challenges and experiences in present scenario.

Simply explained, Water Pollution is an important book for all who wish to make a difference in how to plan and manage our water resources.

Until and unless Water, that magical substance from which all life springs forth, is essential to the very existence of every life form on earth. The role of water in the living organism has not changed since life's first creation in salt water billions of years ago.

Dr. Hemant Pathak

M.Sc. (Gold medalist), Ph. D.

Assistant Professor of Engineering Chemistry

Indira Gandhi Govt. Engineering College,

Sagar, MP, India

Glossary

Aquifers: Underground formations, usually composed of sand, gravel, or permeable rock, capable of storing and yielding significant quantities of water.

Artesian: Describes a confined aquifer containing groundwater that will flow upwards out of a well without the need for pumping.

Catchment area: The area that draws surface runoff from precipitation into a stream or urban storm drain system.

Discharges: Defined by the Clean Water Act as the addition of pollutants (including animal manure or contaminated waters) to navigable waters.

Estuaries: Coastal waters where seawater is measurably diluted with freshwater; a marine ecosystem where freshwater enters the ocean.

Freshwater: Water without significant amounts of dissolved sodium chloride (salt). Characteristic of rain, rivers, ponds, and most lakes.

Groundwater: Water contained in porous strata below the surface of the Earth.

Non-aqueous phased liquids (NAPL): Organic liquids that are relatively insoluble in water and less dense than water. When mixed with water or when an aquifer is contaminated with this class of pollutant (frequently hydrocarbon in nature), these substances tend to float on the surface of the water.

Nonpoint source: A diffuse, unconfined discharge of water from the land to a receiving body of water. When this water contains materials that can potentially damage the receiving stream, the runoff is considered to be a source of pollutants.

Permeability: The ease with which water and other fluids migrate through geological strata or landfill liners.

Point source: An identifiable and confined discharge point for one or more water pollutants, such as a pipe, channel, vessel, or ditch.

Porosity: The total volume of soil, rock, or other material that is occupied by pore spaces. A high porosity does not equate to a high permeability because the pore spaces may be poorly interconnected.

Recharge: A hydrologic process where water moves downward from surface water to groundwater. This process usually occurs in the vadose zone below plant roots, and is often expressed as a flux to the water table surface.

Sorption: The physical or chemical linkage of substances, either by absorption or by adsorption.

Total maximum daily load: The maximal quantity of a particular water pollutant that can be discharged into a water body without violating a water quality standard.

Watershed: The area of land that drains into a lake or stream.

Water Pollution

1.1 Introduction

The quality of water should under no circumstances be degraded to the recommended level from a better level." –The World Health Organization

Today 70% of world population faces serious challenges to get drinkable water and adequate water supply. Although water sources are abundant on earth, about 71% of earth surface is covered with water. Water covers about three-quarters of Earth's surface and is a necessary element for life. These fresh water resources have in many cases been overexploited or severely polluted. Water pollution occurs when a body of water is affected with large amounts of harmful substances. When water is considered unfit for its intended use, it is called polluted. Humans need water for drinking, sanitation, agriculture, and industry; and contaminated water can spread illnesses and disease vectors, so clean water is both an environmental and a public health issue.

Water pollution refers to harmful substances released into surface or ground water, either directly or indirectly. Water is a unique substance, because it can naturally renew and cleanse itself, by allowing pollutants to settle out (through the process of sedimentation) or break down, or by diluting the pollutants to a point where they are not in harmful concentrations.

Besides this Increase, population and its necessities have also lead to the deterioration of surface and sub surface waters. Human never cared for maintain the quality of water; and always destroyed the rivers, lakes, and oceans.

However, this natural process takes time, and is difficult when excessive quantities of harmful contaminants are added to the water. And humans are using more and more

materials that are polluting the water sources that we drink from. Water pollution occurs when energy and other materials are released, degrading the quality of the water for other users. Water pollution includes all of the waste materials that cannot be naturally broken down by water. In other words, anything that is added to the water, above and beyond its capacity to break it down, is pollution.

25% of the human body is made up of solid matter and remaining 75% is water. Therefore, if our bodies are not continuously supplied with water, our bodies become dehydrated and the vital organs will deteriorate until they are no longer viable for human life. Available data indicate that nearly 1.5 billion people lack safe drinking water in world. At least 5 million deaths per year have been attributed only to water borne disease. Thus, human activities and their impact on quality of water resource has been a matter of more concern.

Water also acts as a purifier in our bodies. If enough water isn't consumed, one is unable to properly flush out their kidneys and/or liver, and the colon is unable to expel bowels properly and completely thus keeping unhealthy toxins in the body. As a result, the toxins are able to make its way through the human body causing poisoning and spreading infections.

Water sustains life for humans, animals and plants. People need water for basic everyday activities like drinking and cooking, but water is also very important for the fuelling of agriculture and industry, and plays an important role in the nature of national economies.

However the supply of freshwater available to humanity is shrinking. One of the main causes of this is the polluting of many freshwater resources. In some countries lakes and rivers have become polluted with an assortment of waste, including untreated or partially treated municipal sewage, toxic industrial effluents,

Harmful chemicals and ground waters from agricultural activities. Polluted water supplies not only limit water availability but also put millions at risk of water-related diseases.

The lack of freshwater is likely to be one of the most critical natural resource issues facing people in the next 50 years. The world's population is expanding rapidly, yet our supplies of freshwater are not, placing greater demand on our water resources. This makes it even more important that the remaining freshwater we have is kept safe and clean.

1.2 Water Cycle

Total volume of water in hydrosphere is estimated to be 1.4 billion km^3 of which 97% is ocean water and rest 3% is available as fresh water. During their constant cycling between land, the oceans, and the atmosphere, water molecules pass repeatedly through solid, liquid, and gaseous phases, but the total supply remains fairly constant. Water vapor redistributes energy from the sun around the globe through atmospheric circulation. This happens because water absorbs a lot of energy when it changes its state from liquid to gas. Even though the temperature of the water vapor may not increase when it evaporates from liquid water, this vapor now contains more energy, which is referred to as latent heat.

Atmospheric circulation moves this latent heat around Earth, and when water vapor condenses and produces rain, the latent heat is released.

Three basic steps in the water cycle: water precipitates from the atmosphere, travels on the surface and through groundwater to the oceans, and evaporates or transpires back to the atmosphere from land or evaporates from the oceans.

1.3 Water resources

Water resources are sources of water that are useful or potentially useful to humans. Water resources are under major stress around the world. Rivers, lakes, and underground aquifers supply fresh water for irrigation, drinking, and sanitation, while the oceans

provide habitat for a large share of the planet's food supply. Today, however, expansion of agriculture, damming, diversion, over-use, and pollution threaten these irreplaceable resources in many parts of the globe.

(a) Surface water

Surface water is water in a river, lake or fresh water wetland. Surface water is naturally replenished by precipitation and naturally lost through discharge to the oceans, evaporation, and sub-surface seepage.

(b) Ground water

Groundwater is fresh water located in the pore space of soil and rocks. It is flowing within aquifers. There is a distinction between sub-surface water that is closely associated with surface water and deep sub-surface water in an aquifer. The natural input to sub-surface water is seepage from surface water. The natural outputs from sub-surface water are springs and seepage to the oceans.

1.4 Causes of Water Pollution

Everybody need clean water to survive. Industrialization and urbanization demands, population growth and resultant increased agricultural production needs, and increasing water supply and sanitation needs.

Pollutants that affect our water resources come from many different places. One of the main causes of pollution is the water and other pollutants that flow into storm drains. Storm water is the water that comes from rain and runs off into gutters called storm drains. Urban runoff is water that comes from water used in human activities such as cleaning the car to watering the garden.

Storm water and urban runoff are carried to streams, rivers and eventually to the ocean. This runoff can contain many contaminants that affect clean water.

Few countries have adequately safeguarded water quality and controlled water pollution. Many countries do not have standards to control water pollution adequately, while others cannot enforce water quality standards. The consequence of having polluted water is:

• A reduction in water quality which leads to health problems

• The endangerment of agriculture and aquatic ecosystems.

Pollution of rivers and lakes reduces accessible freshwater supplies. Each year roughly 450 cubic kilometres of wastewater are discharged into rivers, streams and lakes. To dilute and transport this dirty water before it can be used again, another 6,000 cubic kilometres of clean water are needed - an amount equal to about two-thirds of the world's total annual useable fresh water runoff.

1.5 Water Pollution and types

The loss of any of the actual or potential beneficial uses of water caused by any change in its composition due to human activity. Any pollutants that are deposited underground can adversely affect water supplies. The quality of ground water depends on various chemical constituents and their concentration. Different pollutants include:

• Landfill sites and rubbish dumps

• Buried harmful waste e.g. Fuel tank

  • Industrial or mining waste

• Human waste

The beneficial uses of water include its use for drinking, domestic purposes, watering livestock, irrigation of crops, industry, food production, for bathing, for recreational and amenity use. Contamination of water supplies was coming, not only from cities and industry, but also from livestock and field runoff sources.

Water pollution divides into the following categories:

• Point source pollution

Point sources of pollution occur when harmful substances are emitted directly into a body of water, for example an instance where there is an oil spill. Water pollutants that are discharged from specific locations (e.g., factories, wastewater treatment facilities, septic systems).

There can be traced to a specific source, such as a Leaking chemical tank, effluents coming from a waste treatment or industrial plant.

• Non-point source pollution

A non-point source delivers pollutants indirectly through environmental changes, for example when fertilizer from a field is carried into a stream by rain. Non-point sources are more difficult to identify, because they cannot be traced back to a particular location. Water pollutants that come from widespread areas and cannot be tracked to a single point or source or pollutants that are discharged from specific locations including sediment, fertilizer, chemicals and animal wastes from farms, fields, construction sites and mines, are all examples of non-point source pollution.

Landfills can also be a non-point source of pollution, if substances leach from the landfill into water supplies.

• Biodegradable waste pollutants

It consists mainly of human and animal waste. When biodegradable waste enters a water supply, the waste provides an energy source (organic carbon) for bacteria. Organic carbon is converted to carbon dioxide and water, which can cause atmospheric pollution and acid rain; this form of pollution is far more widespread and problematic than other forms of pollutants, such as radioactive waste. If there is a large supply of organic matter in the water, oxygen-consuming (aerobic) bacteria multiply quickly, consume all available oxygen, and kill all aquatic life.

• Nutrients pollution

Plant nutrients, such as phosphates and nitrates, enter the water through sewage, and livestock and fertilizer runoff. Living things cannot survive without nutrients, but too much can be detrimental to watershed organisms. Phosphates and nitrates are also found in industrial wastes. An overabundance of nutrients leads to escalation in plant growth, particularly of algae and vascular plants. Though these chemicals are natural, 80 percent of nitrates and 75 percent of phosphates in water are human-added. When there is too much nitrogen or phosphorus in a water supply (0.3 parts per million for nitrogen and 0.01 parts per million for phosphorus), algae begin to develop.

When algae blooms, the water can turn green and cloudy, feel slimy, and smell bad. Weeds start to grow and bacteria spread. Decomposing plants use up the oxygen in the water, disrupting the aquatic life, reducing biodiversity, and even killing aquatic life. This process, called eutrophication, is a natural process, but generally occurs over thousands of years. Eutrophication allows a lake to age and become more nutrient-rich; without nutrient pollution, this may take 10,000 years, but pollution can make the process occur 100 to 1,000 times faster. Sources of nutrient pollution include overflow from sewage treatment plants, leakage from improperly maintained septic systems, discharge from factories, and automobile exhaust. Examples of nutrient pollutants include fertilizers, animal manure, discharge from boat toilets, and household detergents.

- **Thermal pollution**

Heat can be a source of pollution in water. As the water temperature increases, the amount of dissolved oxygen decreases and raises the risk of disease and death to aquatic organisms. Thermal pollution can be natural, in the case of hot springs and shallow ponds in the summertime, or human-made, through the discharge of water that has been used to cool power plants or other industrial equipment. Fish and plants require certain temperatures and oxygen levels to survive, so thermal pollution often reduces the aquatic life diversity in the water.

- **Sediment pollution**

Sediment is dirt, minerals, sand, and silt eroded from the land and washed into the water. Sediment is one of the most common sources of water pollution. Sediment consists of mineral or organic solid matter that is washed or blown from land into water sources. Sediment can cause large problems, as it can clog municipal water systems, smother aquatic life, and cause water to become increasingly turbid. And, turbid water can cause thermal pollution, because cloudy water absorbs more solar radiation.

Sediment causes several problems for aquatic organisms. First, particles of sediment are suspended in the water. The resulting cloudiness decreases the amount of sunlight that can reach underwater plants that provide food and oxygen for underwater animals. Sediment pollution is difficult to identify, because it comes from non-point sources, such as construction, agricultural and livestock operations, logging, flooding, and city runoff. Sediment particles settle, they fill spaces between rocks, destroying the habitat needed by many underwater insects and other macro invertebrates. Sediment also clogs the gills of fish, crabs, and other underwater organisms.

- **Agricultural pollutants**

Agriculture is the biggest polluter. In virtually every country where agricultural fertilisers and pesticides are used, they have contaminated groundwater aquifers and surface waters. Animal wastes are another source of pollution in some areas. The water that goes back into rivers and streams after being used for irrigation is often severely degraded by excess nutrients, salinity, pathogens, and sediments that it is unfit for further use unless cleaned by water purification plants which are very costly. The water that goes back into rivers and streams after being used for irrigation is often severely degraded by excess nutrients,

salinity, pathogens, and sediments that it is unfit for further use unless cleaned by water purification plants which are very costly.

1. Agricultural pesticides

Pesticides are used as a common name for all chemicals, natural, or synthetic, used to control animal or plant life. Pesticides are chemical compounds toxic to certain living organisms, from bacteria and fungi up to higher plants and even mammals. Pesticides are of great value to agriculture and public health.

Run-off from farms, backyards, and golf courses contain pesticides such as DDT that in turn contaminate the water affect the environment negatively. Most pesticides are compounds, which do not occur naturally in the environment, pesticides are mobile in the soil, and therefore, detectable concentrations indicate pollution. Careless use of pesticides can contaminate water sources and make the water unsuitable for drinking.

The most widely used are insecticides (for extermination of insects), herbicides (for extermination of weeds and other undesirable plants), and fungicides (for preventing fungal diseases). People who use pesticides on their gardens and lawns tend to use ten times more pesticide per acre than a farmer would.

2. Fertilizers

Fertilizers promote growth of algae and aquatic vegetation beyond what is naturally sustainable. This growth reduces water clarity, often covering the entire surface of a pond or lake. Such a "bloom" of algae consumes oxygen, causing fish kills.

Excess nitrate levels from fertilizer can contaminate drinking water supplies. The forms of nitrogen that cause problems as pollutants are the nitrate and ammonium forms.

The nitrate form is water-soluble and moves with the water into surface water or groundwater. The ammonium form attaches to soil particles. It is of less concern for groundwater pollution, but it will wash into surface water if the soil erodes. Both synthetic and natural fertilizers cause nitrate problems if not managed properly.

• Industrial Pollution

Wastes from industries have increased enormously in recent decades. They not only affect freshwater supplies and everything dependent on them, but also marine life. The traditional agriculture-based economies of world are giving way to industrial economies. This transformation is having serious environmental side-effects, particularly in the case of pollution. Efforts have been made to improve regulation, but the absence; Major sources of pollution are industries producing metals, paper and pulp, textiles, and food and beverages. The mining industry is also a significant contributor.

Industrial pollutants, such as wastes from chemical plants, are often dumped directly into waterways while oils and salts are washed off city streets.

A particular group of synthetic chemicals are an immense threat; these are known as persistent organic pollutants (pops):

• DDT- this is an insecticide mainly used to kill mosquitoes, flies, fleas, lice and agricultural pests.

• Dioxins- these chemical compounds occur as a by-product of industrial processes like paper bleaching, and also when plastics are burnt.

• PCB's (polychlorinated biphenols)- chemicals used to manufacture items like transformers, pumps, plastics, paints and adhesives.

They are long-lived and highly toxic in the environment and not break down easily under natural processes. Thus they tend to affect all species up food chain, until they pose risks to human health.

- **Hazardous and toxic chemicals pollution**

Hazardous and toxic chemicals are usually human-made materials that are not used or disposed of properly. Domestic and personal use of chemicals can significantly contribute to chemical pollution.

Water quality has been impacted worldwide by industrial and agriculture chemicals. Eutrophication of surface waters from human agricultural waste and nitrification of ground water from agricultural practices has greatly affected large parts of the world.

Household cleaners, dyes, paints and solvents are also toxic, and can accumulate when poured down drains or flushed down the toilet. It includes chemicals that poison and kill organisms in and near streams, rivers, lakes, and the Bay. When a body of water has a high level of toxic pollution, fishing for the purpose of human consumption is banned.

Even low levels of toxicity can be lethal when chemicals accumulate in predators that consume large amounts of slightly poisoned organisms. Examples of toxic pollution include pesticides and herbicides; gasoline, oil, and other automotive products; household cleaning products; paints and solvents; battery acid; industrial waste chemicals; and toxic substances in car exhaust and solid waste incinerator smoke.

Many of the water that we drink contain toxic chemicals. These chemical substances that are found in water are due to natural processes or human activities. Examples of the most common chemicals found in water include the following:

• **Arsenic:** Arsenic is a naturally occurring metal found in all lead, copper, and gold ores. The Environmental Protection Agency has set the acceptable standard for arsenic at 10 parts per billion in tap water, however many areas exceed this limit. The International Academy for Research on Cancer (IARC) classifies arsenic as a Category I carcinogen. Other health risks at high levels include vascular disease, liver disease, skin lesions, and neurological disorders.

• **Fluoride:** Fluoride is a natural occurring chemical found in foods and water. Fluoride is an essential element due to its importance in bone and tooth structure and growth. However, at high levels, fluoride can be toxic.

• **Chlorine:** Chlorine is often added to water in order to kill bacteria. If consumed in high amounts, chlorine can be toxic and cause sufficient cell damage in the human body.

• **Iodine:** Unlike some other chemicals, insufficient levels of iodine can cause lead to severe health problems such as the enlargement of the thyroid gland, mental retardation, and cretinism. "Water is one of the main sources of dietary intake of iodine.

• **Nitrates:** Pesticides, nitrogenous fertilizers, and manure are the common sources of the presence of nitrates in water. High levels of nitrate in water can lead to blood poisoning and eventually death.

• **Radioactive pollutants**

It includes wastewater discharges from factories, hospitals and uranium mines. These pollutants can also come from natural isotopes, such as radon. Radioactive pollutants can be dangerous, and it takes many years until radioactive substances are no longer considered dangerous.

• Domestic and Industrial sewage

In developing countries statistics suggest that almost all sewerage that is deposited into rivers, lakes and the ocean are untreated. Untreated sewage is a major source of groundwater and surface water pollution in the developing countries.

The organic material that is discharged with municipal waste into the water sources uses substantial oxygen for biological degradation thereby upsetting the ecological balance of rivers and lakes. Sewage also carries microbial pathogens that are the cause of the spread of disease.

This causes significant health risks, as water carrying untreated sewerage is a potential source of waterborne disease. The effects can be far reaching, if the water is used to grow crops that are then eaten uncooked then disease can spread to a whole community.

• Organic pollution

Pollution that arises from organic materials. It is broken down by aerobic (oxygen-consuming) bacteria, which utilize dissolved oxygen in the water. This process lowers the dissolved oxygen content of water.

(a) Water can be contaminated by a number of organic compounds, such as chloroform, gasoline, pesticides, and herbicides from a variety of industrial and agricultural operations or applications. It is unlikely groundwater will suddenly become contaminated, unless a quantity of chemicals is allowed to enter a well or penetrating the aquifer.

Surface water may show great variations in chemical contamination levels due to differences in rainfall, seasonal crop cultivation, and industrial effluent levels. Also, some hydrocarbons (the chlorinated hydrocarbons in particular) form a type of contaminant that is especially troublesome. These are a group of chemicals known as dense non-aqueous phase liquids, or DNAPLs. These include chemicals used in dry cleaning, wood preservation, asphalt operations, machining, and in the production and repair of automobiles, aviation equipment, munitions, and electrical equipment. These substances are heavier than water and they sink quickly into the ground, once the DNAPLs have reached an aquifer they tend to move laterally under the influence of gravity and to slowly dissolve into the groundwater, providing a long-term source for low level contamination of groundwater. Because of their movement patterns DNAPL contamination is difficult to detect, characterize and remediate. All of the pesticides are common neurotoxicants. They acting on the peripheral and/or central nervous systems with the different mechanism. They can accumulate in the food chain from plants to animals this could be interfere or destroy the function of nervous system.

• Inorganic pollution

Inorganic pollutants also readily enter the water system. There are several classes of inorganic pollutants:

(a) Metals

Disease which comes from water pollution is the big problem to human health. Heavy metals represent problems in terms of groundwater pollution.

There are concerns of chronic exposure to low levels of heavy metals in drinking water. Some chemical and heavy metals such as lead, cadmium, and mercury can contaminate and in turn pollutes the surrounding area.

They originate from human activity, especially mining and smelting. Some examples include lead, mercury, tin, and cadmium.

Heavy metals like mercury and certain other water-soluble chemicals are bind tightly to specific sites within the body. There are many disease which are effect to human health after consume the polluted water.

The health effect of others metals contaminated in drinking water is similarly to the health effect from metal pollutants in waste disposal.

(b) Nonmetallic salts

These include selenium and arsenic salts, which have to some extent always been a problem in desert areas. Sodium chloride is the curse of irrigation. They can pass through the porosity of soil and make ground water become polluted. This contamination of groundwater and soil through landfills is known as leaching. Sometime the risk of health effect from contaminated ground water or surface water occurred by the result of bioaccumulation process which is the important process by which chemicals can affect living organisms is through bioaccumulation.

(c) Acids and Bases

Acids are released as by-products of industrial processes. Coal mining releases sulfur compounds, which eventually form sulfuric acid. Coal combustion releases sulfur dioxide into the air, which eventually forms sulfuric acid, falling as acid rain. pH of lakes and streams drops. Inorganic matter high in plant nutrients, such as phosphate and nitrate compounds, encourages rapid plant growth. When these plants eventually die and decay, they are decomposed by aerobic bacteria, and dissolved oxygen levels decrease as a result.

High levels of some of these chemicals can be directly dangerous to human health. Some, including heavy metals and salts, can render water highly hazardous to human health and reduce ecosystem functions.

• Oil pollution

In fact, one drop of used motor oil can pollute 25 litres of water. This arises from accidental or intentional releases of oil during the recovery, transport, and use of oil. Oil floats on the surface and oil spill can be disastrous for marine life. The oil film can stop oxygen dissolving into the water and can coat bird feathers or seal fur.

Oil pollution causes immediate death to many organisms and contaminates many others. Cleanup is extraordinarily expensive and often ineffective.

• Bacterial pollution

It occurs when there is an excess of harmful bacteria. There are many beneficial bacteria in the water. Even harmful bacteria in small amounts are safe. In larger concentrations, however, certain types of bacteria can be deadly to fish and animals (including humans) that drink or accidentally ingest the water.

Certain bacteria can cause illness if they come in contact with an open wound. Interestingly, most of these harmful bacteria do not affect aquatic insects. Some sources of bacterial pollution include overflow from sewage treatment plants, leakage from improperly maintained septic systems, animal manure, and discharge from boat toilets.

1.6 Water-Related Diseases

More than 2 million people die each year from diseases such as cholera, typhoid, and dysentery that are spread by contaminated water or by a lack of water for hygiene. Turbidity (cloudiness) in drinking water, an indicator of possible contamination.

It is believed that many of the diseases associated with water contamination are caused by pathogens. Pathogens spread by water leading to the spread of diseases. Examples of pathogens are bacteria, viruses, and parasites. Pathogens are considered to be communicable because they have the ability to spread from one person to another by way of contaminated water and/or other vectors." The most common diseases spread through water are diarrheal diseases; examples include cholera, typhoid, paratyphoid, salmonella, giardiasis, and cryptosporidiosis.

E.Coli has been identified as a major indicator fecal organism present in drinking water. According to standards set by the World Health Organization, no drinking water should contain E.Coli in any 100 mL sample.

Many pathogens arise from animal and human feces and from insufficient water supply. Pathogen related diseases could also be spread by person-to-person contact, aerosols, and food intake.

This causes the bacteria causing the disease to buildup over time causing an infected person to contaminate the water supply and therefore continuing the spread of disease. Water-Related Diseases classified in to four categories:

• **Waterborne Diseases:** Waterborne diseases arise from the contamination of water by human and/or animal body excretions infected by pathogenic viruses or bacteria, which are directly transmitted when the water is consumed or used for food preparations. Examples include cholera, typhoid, and cryptosporidiosis.

• **Water-Privation Diseases:** Water-Privation diseases are affected by the quantity of water. The disease is spread through (infected) person-to-person contact or through contact with infected materials. Poor personal hygiene is a common factor leading to water-privation diseases.

• **Water-Based Diseases:** In water-based diseases, water provides the habitat for intermediate host organisms in which parasites are able to pass part of their life cycle and later their infective larval forms in water are passed on the humans.

• **Water-Related Diseases:** In water-related diseases, water provides a home for insects. Examples of these types of diseases include malaria, dengue and yellow fever.

• **Water-Dispersed Infections:** In these types of infections, pathogens are able to Proliferate in freshwater and enter the body through the respiratory tract.

1.7 Importance of Water to Human Health and Survival

Water is susceptible to various toxic pathogens and chemicals. Therefore, it is of extreme importance that the quality of water is tested through frequent monitoring.

The quality of drinking water can be determined by its appearance, taste, and odor. Fecal organisms have been deemed as a popular organism to monitor in water because of its ability to be present in high numbers in the feces of humans and animals and because of its ability to be easily detected.

Water sustains life for humans, animals and plants. People need water for basic everyday activities like drinking and cooking, but water is also very important for the fuelling of agriculture and industry, and plays an important role in the nature of national economies.

Water resources are coming under intense pressure by population growth and the need for increased agricultural production. Inadequate provision of sanitation facilities, sewerage and wastewater treatment results in significant quantities of this wastewater reaching water bodies that may service human consumption.

However the supply of freshwater available to humanity is shrinking. One of the main causes of this is the polluting of many freshwater resources. In some countries lakes and rivers have become polluted with an assortment of waste, including untreated or partially treated municipal sewage, toxic industrial effluents, harmful chemicals, and ground waters from agricultural activities.

Polluted water supplies not only limit water availability but also put millions at risk of water-related diseases.

The lack of freshwater is likely to be one of the most critical natural resource issues facing people in the next 60 years. The world's population is expanding rapidly, yet our supplies of freshwater are not, placing greater demand on our water resources. This makes it even more important that the remaining freshwater we have is kept safe and clean.

1.8 Managing Water Quality

Several countries are implementing ambitious programmes to build wastewater treatment plants and rehabilitate degraded water resources. Examples include China, India, Thailand, the Philippines, Bangladesh, and Indonesia. These and many more have passed water quality acts or laws to prevent pollution and protect receiving waters. Unfortunately enforcement is challenging, especially in emerging economies where institutional capacities cannot keep pace with rapid industrialization, and economic instruments like taxation and removal of fertilizer subsidies clash with development goals. Monitoring is also costly and voluntary compliance unlikely.

The transboundary nature of many river basins, and the need for their collaborative management, improved and effective water quality management strategies in Asia require the collection, analysis, and sharing of accurate data. Currently this task is, with some exceptions, generally poorly implemented. In most countries sporadic or patchy data collection prevails, and it is often accompanied by inadequate analysis.

• Water management in India

Water management in India in future must shift its emphasis from social good as at present to economic good, and use of market mechanisms with the participation of the business sector. Such a change will achieve a more efficient and effective allocation, use and management of water, and their roles of both public and private sectors in managing water resources must be defined.

1.9 Drinking water: A challenge for world population

"Access to safe water is a fundamental human need and, therefore, a basic human right. Contaminated water jeopardizes both the physical and social health of all people. It is an affront to human dignity." -Kofi Annan, UN Secretary-General

World population to face serious water quality issues that contribute to freshwater scarcity, ill-health, and even deaths. In many places quality is continuing to decline and insufficient efforts are being made to monitor and remedy the situation amid institutional and social challenges. However, there are also robust efforts to correct the situation and cause to be hopeful.

Our drinking water resources is not safe, contaminated with toxic pathogens and chemicals; Sources of water contaminants include human and animal fecal wastes, improper disposal of chemicals, natural occurring floods and other disastrous events, and use of agricultural products such as pesticides and fertilizers.

Water becomes contaminated when it comes in contact with these toxic pathogens and chemicals. The most common type of water contamination is through human and animal feces. The feces enter the water supply and spread through the population through person-to-person contact. Infections caused by pathogens are diarrheal diseases such as E. coli, giardia and the typhoid fever. Some populations are more susceptible to water-contamination diseases more than other populations. For example, children and infants are vulnerable to pathogen related diseases because their immune systems are not fully developed and strong enough to fight off the toxic contaminants and the resulting infections. Statistics show that more than 2 million children die from diarrheal diseases each year with 90% of these children being under 5 years of age. Other populations that are prone to diarrheal diseases include cancer patients, HIV/AIDS patients, transplant patients, the elderly and pregnant women (including their unborn child).

Chemical toxins, pesticides, and fertilizers are also sources that contaminate drinking water. In underdeveloped countries, such as Bangladesh, India, China, Taiwan, and Nepal, people drink water from "arsenic-laced" wells on a daily basis. Examples of health risks that are caused by chemical contamination of drinking water include skin lesions, vascular and cardiac problems, and cancer of the bladder, lungs, or skin, liver and kidney damage, damage to the nervous system, suppression of the immune system, and birth defects.

1.10 Impact of water pollution on the drinking water supply

Poor quality of surface and groundwater has become a threat to worldwide supplies of drinking water. In industrialized regions, excessive nitrate spread over farmland, bacteria, hazardous liquid waste and trace chemicals pose an increasing threat to drinking water supplies. The drinking water supply is more extensively affected by sewage influx, faecal Contamination, pesticides, nitrates and industrial discharges threatening with public health risks. The increasing contamination of groundwater due to persistent types of pollutants that are not infiltrated by the soil is of particular relevance to global water security. Human activities may have impact on water bodies due to disposal of waste. The waste could be due to the spillage of chemical materials, oils and greases. Contamination of ground water can take place, if the dump containing above substances is leached and percolate into the ground water level. Oil spillage during change of lubricants, cleaning and repair processes, in the maintenance of rolling stock, is very common.

(1) It has been estimated that as much as 25% of fertilizers is carried away as run off.

(2) Fertilizer run off has contaminated ground water and polluted bodies of water near and around farmlands.

(3) High and unsafe nitrate concentrations in drinking water have been reported in countries that practice intense farming.

(4) Accumulations of nitrogen and phosphorous in water ways from chemical fertilizers has also contributed to the eutrophication of lakes and ponds.

(5) Ammonia released from the decay of fertilizers causes minor irritation to the respiratory system.

The health of rivers, lakes, estuaries, coastal systems as well as marine resources is threatened world-wide by water pollution issues, such as eutrophication, toxics (pesticides, POPs), heavy metals, acidification and siltation. Their main effects are ecosystem dysfunction, loss of biological diversity, alteration of aquatic habitats and contamination of downstream and marine ecosystems and are particularly acute near centres of human activities.

1.11 Economic impact of water pollution

There is real and potential loss of development opportunity because of diversion of funds for the remediation of water pollution in several developing countries. If remediation costs exceed economic benefits, lending institutions may regard development projects as no longer being creditworthy. In this context, the following message needs to be delivered to decision-makers: the cost of water pollution is higher than the cost of its prevention, and neglecting water pollution control entails high social and environmental costs. Food resources are also threatened by a damaged agricultural production, in terms of decreased crop yield and quality, with salinised and polluted water for crop irrigation.

Aquatic ecosystems will not be able to provide the essential goods either. The damaging of commercial fisheries affects self-sufficient fishing communities and riparian settlements. The decline of commercial fish production is in turn expected to exacerbate demands for protein from livestock production and agriculture.

In some parts of the world, water has also been judged unsuitable even for industrial purposes. This can have a significant impact on industrial productivity and a respective impact on the economy of industrialized or rapidly industrializing countries. The profitable tourism "industry" is also negatively affected by water pollution and consequent esthetical degradation.

1.12 Impact of water pollution on human health and social security

3-4 million people die each year of waterborne diseases worldwide, including more than 2 million children who die from diarrhea. Urban populations in developing countries and particularly in urban slums are groups especially vulnerable to the negative health impact of water pollutants. The costs on human health protection and preventative medicine are significant.

The degradation of water resources reduces social security. The impairment of water resources in regions where poverty already affects a great part of the population, can lead to greater social inequity and poverty intensification. Poor quality water also forces women, who are the main collectors of water, to travel long distances in order to obtain clean water, thus negatively affecting their time management. As far as regional and international security is concerned, degradation and lack of respect for water resources may exacerbate social conflict. Conflicts between upstream and downstream nations or communities shall increase.

1.13 Monitoring Drinking Water

The most effective way to protect the quality of drinking water is through consistent and constant monitoring of the drinking water supply. Before distributing water, it is important to verify that the quality of water meets the standards of safe drinking water. It is also important to ensure that there is sufficient supply of water to meet the demands of the population. Finally, it is important to know where the water is coming from before distribution. For example, if the water is coming from a well, one should verify the location and construction of the well; making sure that the well and its water is protected from surface drainage and flooding. It is also important to keep any source of drinking water isolated from human activities and garbage.

1.14 Treating Drinking Water

Proper treatment and handling of drinking water is essential not only for the quality of water but also for human health. The type of treatment depends on several factors: the quality of the original source of water, the number of people to whom the water will be served, and the point of origin for the water.

For drinking water that does not require much treatment or will be served to a small population (such as to individuals within the same household), water can be treated by water purification filters and disinfecting tablets. Water can also be boiled to remove pathogens—studies show that boiling water for one minute can remove most or all pathogenic contaminants. However, there have been warnings placed for boiling water containing lead and nitrate. Boiling water that contains lead and nitrate actually increases its concentration and therefore poses as a greater risk for infection and poisoning. It is important to keep in mind that boiling water cannot be used to remove chemical toxins. Boiling water can also be expensive, especially in places where fuel is limited and the population is high.

In water distribution centers, different methods are used to treat drinking water: flocculation, sedimentation, filtration, and disinfection:

• **Flocculation:** The process of flocculation involves removing dirt and other particles that are suspended in the water. This is achieved by adding aluminum and iron salts to the water; these salts form sticky particles, which attract the particles to be removed.

• **Sedimentation:** Once the particles are removed by flocculation, they naturally settle out of the water.

• **Filtration:** Filtration is used to remove particles such as clays, organic matter and, chemicals, and precipitates out of the water. The process of filtration purifies the quality of drinking water and thus decreasing the chances of contamination.

• **Disinfection:** The process of disinfection is one of the most popular and advanced treatment methods of the 20th century. Disinfection occurs before water has a chance to enter the distribution center to ensure that the water is free from toxins. Effective and efficient sources of disinfectants include chlorine, chlorinates, and chlorine dioxides.

According to researchers, not all organisms and contaminants found in drinking water are harmful as long as they do not exceed safety values established by the World Health Organization. Therefore, contaminants should be removed when absolutely necessary. It is important to keep in mind that removing contaminants is costly; the more toxins that need to be removed, the more expensive the cost will be. It is also important to remember that removing contaminants does not secure increased safety of human health.

1.16 Prevention of Water Pollution

Water pollution can be controlled by correctly dispose of hazardous household products. Avoid letting contaminated water such as chemicals, soaps, grass clippings, paint etc run into storm drains. Scientists are constantly researching how various contaminants affect drinking water and how these contaminants affect individuals. They continuously study how toxic substances affect a community in order to determine the relationship between exposure to a contaminant and a health effect. Through research and investigation, scientists are also able to determine at what level is a pathogen and chemical toxic or not.

Recycle and dispose of all rubbish properly. Ensure that litter is thrown in the rubbish bin and does not get blown away. Conserve water in the home and garden. Use efficient plumbing fixtures so only the necessary amount of water is used for flushing toilets and showers. Use less water when washing cars. Use a bucket with water and soap to wash the car, and then drain the dirty/soapy water down the sink or in the grass. Only use the hose when it is needed – do not let it run constantly. Use natural fertilizers in the garden. Acertain the maintenance of the car so it does not leak oil and release bad fumes from the exhaust. Volunteer for a beach clean up, tree planting or water quality monitoring. Conserve water; the less water you use, the less will be running down the drains and into gutters, carrying pollutants with it. For more information about water consumption, as well as some tips on how to conserve water, see the Water Consumption fact sheet. Keep pet litter and debris out of street gutters. Use pesticides sparingly; in general, people tend to use 10 to 50 times more fertilizer on their lawns and gardens than is necessary for good plant health. Or, use compost to fertilize your garden. Keep your vehicles running properly. If you have an oil leak, fix it immediately, and if you change your own oil, dispose of the used oil properly. Use natural cleaners, such as baking soda, vinegar and borax. Use detergents with less phosphate; sewage plants can only remove about 30 percent of the phosphates from waste.

1.17 Conclusions and recommendations

Water is the essence of basic survival. Without it, life on Earth would cease to exist. In order to ensure that human life continues to exist, we must work together and do our part to improve the quality of drinking water.

Overuse and pollution of the world's freshwater resources are a recent development. Their long-term consequences are still unknown. Already, however, they have taken a heavy toll on the environment, and they pose increasing risks for
many species.

Researchers must do their part in the laboratory to come up with treatment methods to improve the quality of drinking water. When the quality of drinking water is good, human health is also.

Many chemical substances emitted into the environment from anthropogenic sources pose a threat to the functioning of aquatic ecosystems and to the use of water for various purposes.

The need for strengthened measures to prevent and to control the release of these substances into the aquatic environment has led many countries to develop and to implement water management policies and strategies based on, amongst others, water quality criteria and objectives.

To provide further guidance for the elaboration of water quality criteria and water quality objectives for inland surface waters, and to strengthen international co-operation the following recommendations have been put forward (UNECE, 1993):

• The precautionary principle should be applied when selecting water quality parameters and establishing water quality criteria to protect and maintain individual uses of waters.

• In setting water quality criteria, particular attention should be paid to safeguarding sources of drinking-water supply. In addition, the aim should be to protect the integrity of aquatic ecosystems and to incorporate specific requirements for sensitive and specially protected waters and their associated environment, such as wetland areas and the surrounding areas of surface waters which serve as sources of food and as habitats for various species of flora and fauna.

• Water-management authorities in consultation with industries, municipalities, farmers' associations, the general public and others should agree on the water uses in a catchment area that are to be protected. Use categories, such as drinking-water supply, irrigation, livestock watering, fisheries, leisure activities, amenities, maintenance of aquatic life and the protection of the integrity of aquatic ecosystems, should be considered wherever applicable.

• Water-management authorities should be required to take appropriate advice from health authorities in order to ensure that water quality objectives are appropriate for protecting human health.

• Under no circumstances should the setting of water quality objectives (or modification thereof to account for site-specific factors) lead to the deterioration of existing water quality.

• In setting water quality objectives for a given water body, both the water quality requirements for uses of the relevant water body, as well as downstream uses, should be taken into account. In transboundary waters, water quality objectives should take into account water quality requirements in the relevant catchment area. As far as possible, water quality requirements for water uses in the whole catchment area should be considered.

• Water quality objectives for multipurpose uses of water should be set at a level that provides for the protection of the most sensitive use of a water body. Among all identified water uses, the most stringent water quality criterion for a given water quality variables should be adopted as a water quality objective.

• Established water quality objectives should be considered as the ultimate goal or target value indicating a negligible risk of adverse effects on use of the water and on the ecological functions of waters.

• The setting of water quality objectives should be accompanied by the development of a time schedule for compliance with the objectives that takes into account action which is technically and financially feasible and legally implementable.

• The setting of emission limits on the basis of best available technology, the use of best environmental practices and the use of water quality objectives as integrated instruments of prevention, control and reduction of water pollution, should be applied in an action-oriented way. Action plans covering point and diffuse pollution sources should be designed, that permit a step-by-step approach to water pollution control which are both technically and financially feasible.

Where necessary, a step by- step approach should be taken to attain water quality objectives, making allowance for the available technical and financial means for pollution prevention, control and reduction, as well as the urgency of control measures.

• The public should be kept informed about water quality objectives that have been established and about measures taken to attain these objectives.

Polluted water and lack of sanitation also greatly risk human health. Moreover, the state of freshwater resources contributes to the deterioration of coastal waters and seas.

It is therefore critical that more care is taken to reduce pollutants in our fast retreating freshwater supplies.

1.17 References

1. Hemant Pathak, Pollumeter: A Water Quality Index model for the assessment of water quality in Sagar city, M.P., India. The Green pages: Directory for Environmental Technology, 2011. http://www.eco-web.com/edi/110128.html

2. Hemant Pathak, Study of seasonal variation in ground Water quality Chemical parameters of Sagar city (M.P.) by principal component analysis and evaluation, vol. 8(4), E- Journal of chemistry,2011, ISSN: 0973-4945, www.ejchem.net/PDF/V8N4/2000-2009.pdf

3. Hemant Pathak, Interdependency between physicochemical water pollution indicators: a case study of river Babus, Sagar, M.P., India. Analele UniversităŃii din Oradea – Seria Geografie, Year XXI, no. 1/2011 (June), ISSN 1454-2749, E-ISSN 2065-1619, http://istgeorelint.uoradea.ro/Reviste/Anale /Art/...1/02_ AUOG _515_Hemant.pdf

4. Hemant Pathak, A mathematical modeling with respect to DO for environmentally contaminated drinking water sources of Sagar city (M.P.), India: A case study, Ovidius University Annals of Chemistry, Vol. 22(2), 2011. ISSN-1223-7221, www.univ-ovidius.ro/anale-chimie/chemistry/2011-2/2_pathak.pdf

5. Hemant Pathak, Seasonal study with interpretation of the chemical characteristics of water pond in reference to quality assessment: A case study, Analele UniversităŃii din Oradea – Scria Geografie, vol. 2/2011 (Dec.), ISSN 1454-2749, E-ISSN 2065-1619.

6. Hemant Pathak, Assessment of Physico-Chemical Quality of Groundwater in rural area nearby Sagar city, MP, India, Advances in Applied Science Research (Pelagia Research Library) , 2012, vol. 3 (1), pp. 555-562, ISSN: 0976-8610

7. Hemant Pathak, Studies on the physico-chemical status of two water bodies at Sagar city under anthropogenic Influences, Advances in Applied Science Research (Pelagia Research Library), 2012, vol. 3 (1), pp. 31-44, ISSN: 0976-8610

8. Hemant Pathak, Multivariate evaluation of fluoride contamination in ground water samples of Sagar city, M.P., India: A case study, Instasci Journal of Chemistry, 2012, 2(1), ISSN: 2277-6931

9. Hemant Pathak, Ground and Tap water Quality assessment of Sagar city especially in terms of saturation index, *THE POLYTECHNIC INSTITUTE OF IAŞI,* 2012, Issue LVII (LXI), Fasc. 4. ISSN: 0254 – 7104

10. Hemant Pathak, An water quality index mathematical modeling of water samples of Rajghat, water supply reservoir Sagar (M.P.) with respect to total dissolved solids: A regression analysis, *THE POLYTECHNIC INSTITUTE OF IAŞI,* 2012, Vol. 1, 2012, ISSN: 0254 – 7104

11. Hemant Pathak, Assessment of Physico-Chemical Quality of municipal water samples of Makronia sub-urban area of Bundel khand region, India, Analele UniversităŃii din Oradea – Seria Geografie, vol. 1/2012 (May), ISSN 1454-2749, E-ISSN 2065-1619.

12. Hemant Pathak, Assessment of Physico-Chemical Quality of Groundwater by Multivariate Analysis in some Populated Villages nearby Sagar City, MP, India, J Environ Anal Toxicol 2012, vol. 2(5), ISSN:2161-0525, http://dx.doi.org/10.4172/2161-0525.1000144

13. Hemant Pathak, Eutrophication: Impact of Excess Nutrient Status in Lake Water Ecosystem, Journal of Environment and Analytical Toxicology, 2012, vol. 2(6). ISSN:2161-0525, http://dx.doi.org/10.4172/2161-0525.1000148

14. Hemant Pathak, Indicators of the deteriorate water quality status of reservoir, Sagar city, MP, India by multivariate analysis, Ovidius University Annals of Chemistry, Vol. 23(2), 2012. ISSN-1223-7221.

15. Hemant Pathak, Evaluation of ground water quality using multiple linear regression and mathematical equation modeling, Analele UniversităŃii din Oradea – Seria Geografie, vol. 2/2012, ISSN 1454-2749, E-ISSN 2065-1619.

16. Hemant Pathak, Water quality Studies of two Rivers at bundelkhand region, MP, India: A case study, (accepted by) U.P.B. Sci. Bull., Series, 2012, ISSN 1454-2331

17. Hemant Pathak, A mathematical modeling for environmentally polluted water soluble impurities: A case study(Proceedings of the 45th Annual convention of chemists and international conference on recent advances in chemistry, Organised by, Indian chemical society Hosted by, Karnataka university , Dharwad November 23-27,2008)

18. Hemant Pathak, Assessment of Physico-Chemical Quality of municipal water samples of Sagar city, MP, India (National Seminar on Soil, Air and Water Resource Management, Organised by, Govt. Auto. Girls P.G. College, Sagar, Sponsored by, U.G.C., New Delhi)

19. Hemant Pathak, Contaminant evaluation of boreholes water (drinking water sources) in Gambhiria village, Sagar (MP), April 2012, National symposium on Advances in Environmental chemistry and green technologies, Organised by RJIT Tekanpur, Gwalior, Sponsored by, M.P. Council science and technology, Bhopal

20. Hemant Pathak, Evaluation of ground water quality using multiple linear regression and mathematical equation modeling, September 2012, National conference on Recent Advances in Chemical Sciences; Emphasis on Healthy Life, Organized by ITM, Gwalior, Sponsored by, M.P. Council science and technology, Bhopal

21. Hemant Pathak, Hydrological modeling of waste water discharges in Sagar Lake, September 2012, National conference on Recent Advances in Chemical Sciences; Emphasis on Healthy Life, Organized by ITM, Gwalior, Sponsored by, M.P. Council science and technology, Bhopal

22. Hemant Pathak, Physico-chemical Analysis of Ground Water Samples of Sagar city with respect to water soluble pollutants (National Seminar of Environment protection & waste management , Organised by, Govt. P.G. College, Bina, Sponsored by, U.G.C., New Delhi)

23. Hemant Pathak, Statistical Study on Physico-Chemical Parameters and Water Quality assessment of Lakha banzara pond, Sagar (M.P.) (National Seminar of

Environment Safety and solid waste management, Organised by, Govt. Arts and Commerce College, Sagar, Sponsored by, U.G.C., New Delhi)

24. Hemant Pathak, Assessment of Physico-Chemical Quality of municipal water samples of Sagar city, MP, India (National Seminar on Soil, Air and Water Resource Management, Organised by, Govt. Auto. Girls P.G. College, Sagar, Sponsored by, U.G.C., New Delhi)

25. Hemant Pathak, Evaluation of Water Quality: Physico – Chemical Characteristics of Rajakhedi Village at Sagar city (MP) by using Correlation Study, National Seminar on Recent trends in Environmental sciences, 5th june 2012, Organised by, NRI college of engineering and management, Gwalior (MP).

26. Hemant Pathak, Interpretation of Ground Water Quality Data Variation in Sanodha Village at Sagar city (MP) by using PCA Study, (National Seminar on Recent trends in Environmental sciences, 5th june 2012, Organised by, NRI college of engineering and management, Gwalior (MP).

27. De Kok, T., Guidotti, T., Kjellstrom, T., and Yassi, A. (2001) Basic Environmental Health. New York: World Health Organization.

28. Environmental Protection Agency. Water on Tap: What You Need to Know Global Healing Center. The Toxins In Our Drinking Water: http://www.globalhealingcenter.com/water-toxins.html

30. Grantham, D., Hairston, J., & McFarland, M. (2010) National Water Program: Drinking Water and Human Health: http://www.usawaterquality.org/themes/health/

Herman, G., and Zaslow, S., (1996). Health Effects of Drinking Water Contaminants: http://www.bae.ncsu.edu/programs/extension/publicat/wqwm/he9.html

33. National Resources Defense Council. (2008). Bringing Safe Water to the World: http://www.nrdc.org/international/safewater.asp